Wildflowers of Hells Canyon

ISBN 978-1-9565300-5-6   paperback
ISBN 978-1-9565300-6-3   ebook

Published by Lucky Marmot Press in Wallowa, Oregon.
https://www.luckymarmotpress.com

Wildflowers of Hells Canyon

Janet Hohmann

Lucky Marmot Press

Wallowa, Oregon

Preface

This book contains a small sampling of the more showy species of plants found in Hells Canyon. It is intended as a help to the visitor wanting to identify plants without getting bogged down with too much information. For those who do want more information, please refer to the bibliography.

The book is arranged alphabetically by plant family, found on the upper right corner of each page. The common name and botanical name (genus and species) appear in the upper left page corner. Where the exact species was not known, the botanical name is abbreviated *Genus sp.* Below the botanical name are notations regarding bloom color and whether the plant is native or non-native and invasive. An invasive plant is one which "takes over" and often displaces native vegetation.

Taxonomy is based on *Flora of the Pacific Northwest*, 2nd Edition, by Hitchcock and Cronquist. The Table of Contents lists plants by family and Latin name, and the Index lists plants by common name and by color.

Acknowledgements

Most writing projects happen through the help of many people and institutions, and this book is no exception. I can't begin to name all of the people who have contributed, either directly or indirectly, to the creation of this book. I thank all of you collectively.

In particular, I'd like to acknowledge help from Jenner Hanni, tech wizard extraordinaire, who was essential in scanning hundreds of slides and formatting the manuscript; Holly Goebel, director of the Wallowa Public Library, for making the library space, scanner, and computer available for me to use; and the late Frank Conley, for the use of his slides to supplement my own. Rob Taylor and Kendrick Moholt helped confirm taxonomy.

Bibliography & Further Reading

I have found these plant identification books particularly useful.

1. Craighead, John J., and Frank C. Craighead, Jr, and Ray J. Davis, *A Field Guide to Rocky Mountain Wildflowers.* Houghton-Mifflin Co. 1963.

2. Hitchcock, C. Leo, and Arthur Cronquist. *Flora of the Pacific Northwest: An Illustrated Manual.* University of Washington Press. 1973. 2nd Edition. 2018.

3. Mason, Georgia. *Guide to the Plants of the Wallowa Mountains of Northeastern Oregon.* University of Oregon Press. 1975.

4. Taylor, Ronald J. *Sagebrush Country: A Wildflower Sanctuary.* Mountain Press Publishing Co. 1992.

5. Turner, Marie, and Phyllis Gustafson. *Wildflowers of the Pacific Northwest.* Timber Press, Inc. 2006.

6. Wilson, Ron D. (Editor), and Larry C. Burrell, et. al. *Weeds of the West.* University of Wyoming. 9th Edition. 2000.

.

Table of Contents

This collection is dedicated to my mother,
who passed on to me her love of wildflowers
and the places where they grow.

Allium acuminatum
Tapertip Onion

Amaryllidaceae
Amaryllis Family

Native • Pink / White

We have several species of onions in the canyon.
This is one of the most common. 4-6" tall. Its smell
and taste are easily recognizable as onion. Edible.

Allium tolmeii
Tolmie's Onion

Amaryllidaceae
Amaryllis Family

Native • Pink / White

Another very common species of onion, with
the same pungent smell and taste. Edible.

Toxicodendron radicans
Poison Ivy

Anacardiaceae
Sumac Family

Native • Cream

A smooth-stemmed, 1-4' tall shrub. Leaves
absent in winter. Cream-colored flowers;
greenish-yellow berries. Very common, especially
along waterways. Can cause mild to very severe
itching and/or blistering. If you only learn one
Hells Canyon plant, learn this one well and avoid it!

Lomatium cous
Cous Biscuitroot

Apiaceae
Parsley Family

Native • Yellow

This very small plant blooms in early spring and grows 3-5" tall. Its roots were dug by the Nez Perce, dried and pounded to a flour, and used to make a food that early white settlers thought looked like biscuits, hence the common name.

Lomatium grayii
Gray's Desert Parsley

Apiaceae
Parsley Family

Native • Yellow

This large biscuitroot has finely-dissected,
gray-green leaves. 2-3' tall.

Pterideridea gairdneri
Yampah

Apiaceae
Parsley Family

Native • White

Blooms in early summer in dry, open areas.
Roots used by Nez Perce for food. 12-20" tall.

Asclepias speciosa
Showy Milkweed

Apocynaceae
Dogbane Family

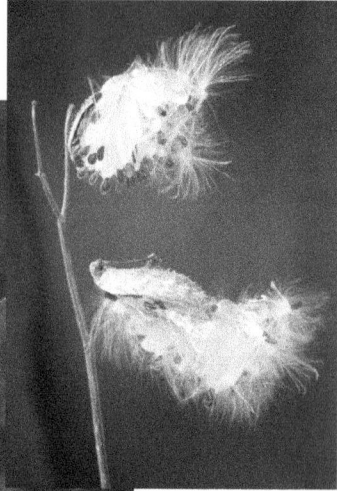

Native • Pink

Seed Pods

This host plant for Monarch butterfly caterpillars
blooms in summer. Milky sap. 2-4' tall.

Camassia cusickii
Cusick's Camas

Asparagaceae
Asparagus Family

Native • Purple

This 24-30" tall camas especially likes the rocky
hillsides along the Hells Canyon Scenic Byway Drive.

Camassia quamash
Common Camas

Asparagaceae
Asparagus Family

Native • Purple

This much-smaller cousin of Cusick's camas grows only 10-12" tall in wet meadows. It was one of the primary roots gathered for food by the Nez Perce.

Tritelia grandiflora
Blue Dicks / Brodiaea / Wild Hyacinth

Asparagaceae
Asparagus Family

Native • Purple

Distinctive ball-shaped flower cluster at the
top of a single, leafless stem. 12-20" tall.

Achillea millefolium
Yarrow

Asteraceae
Daisy Family

Native • White

Blooms in summer. Used medicinally by the Nez
Perce and by settlers. Finely divided foliage. 1-2' tall.

Balsamorhiza incana
Hoary Balsamroot

Asteraceae
Daisy Family

Native • Yellow

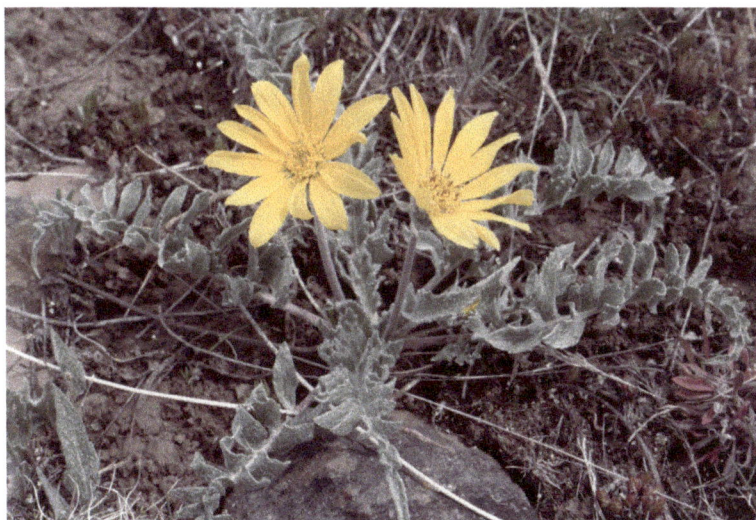

Spring-summer bloom. Note the very jagged
leaves. Hybridizes with *B. sagittata*. Seeds used
as food by Nez Perce. A robust daisy, 1-2' tall.

Balsamorhiza sagittata
Arrowleaf Balsamroot

Asteraceae
Daisy Family

Native • Yellow

Spring-summer bloom. Hybridizes with *B. incana*. Leaves
have smooth margins. Seeds used as food by Nez Perce.

Centaurea maculosa
Diffuse Knapweed

Asteraceae
Daisy Family

Non-Native and Invasive • Pink / White

Common invasive biennial weed. There are several
similar-looking species, all of which grow to 3' tall.

Chaenactis douglasii
Chaenactis / False Yarrow

Asteraceae
Daisy Family

Native • Pink

Distinctive gray-green foliage. 10-20" tall.

Cirsium vulgare
Bull Thistle

Asteraceae
Daisy Family

Non-Native and Invasive • Purple

Fairly common invasive biennial thistle. Can grow
up to 4' tall. Hells Canyon has several species
of thistle, some native and some non-native.

Ericameria nauseosa
Gray Rabbitbrush

Asteraceae
Daisy Family

Native • Yellow

Blooms late-summer to fall. A common shrub, 2-4' tall.

Erigeron sp.
Fleabane Daisy

Asteraceae
Daisy Family

Native • Purple / Yellow

A huge and variable genus; this is one of many *Erigeron* species found in Hells Canyon. Usually 6-15" tall.

Eriophyllum lanatum
Oregon Sunshine / Woolly Yellow Daisy

Asteraceae
Daisy Family

Native • Yellow

Spring-summer bloom. Ray-flowers are
orangish towards central disc, paler yellow
on ends. Grows in clumps, 8-15" tall.

Gaillardia artistata
Gaillardia / Blanket Flower

Asteraceae
Daisy Family

Native • Yellow

Yellow ray flowers may be reddish
towards central disc. 12-20" tall.

Grindelia sp.
Gumweed

Asteraceae
Daisy Family

Native • Yellow

Base of flower very sticky with sap. 12-20" tall.

Haplopappus carthamoides
Golden Weed

Asteraceae
Daisy Family

Native • Yellow

Pale yellow bloom. 12-15" tall.

Helianthella uniflora
Little Sunflower

Asteraceae
Daisy Family

Native • Yellow

Photo by Frank Conley

Another of many sunflower look-alikes.
A single bloom atop a 20-30" stem.

Helianthus annuus
Western Sunflower

Asteraceae
Daisy Family

Native • Yellow

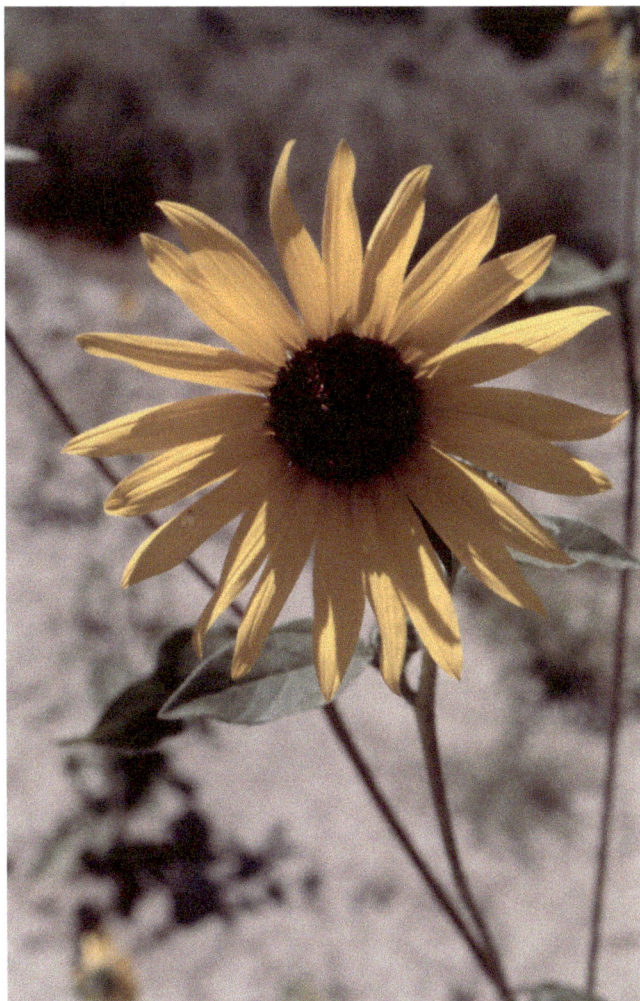

Flower heads face the sun throughout
the day. Grows up to 5' tall.

Matricaria discoidea
Dog Fennel / Pineapple Weed

Asteraceae
Daisy Family

Non-Native and Invasive • White

Photo by Frank Conley

Usually low-growing in disturbed areas. Plant
smells like pineapple when crushed. 6-10" tall.

Microseris sp.
False Dandelion

Native • Yellow

Told from true dandelion by its
long, skinny leaves. 8-12" tall.

Onopordum acanthium
Scotch Thistle

Asteraceae
Daisy Family

Non-Native and Invasive • Purple

This very invasive biennial weed is found throughout
the canyon. In its first year, it produces a flat rosette
of gray-green, spiny leaves. It sends up a 3-5' tall
stalk with light purple blossoms in its second year.

Senecio serra
Groundsel / Butterweed

Native • Yellow

Clusters of yellow, daisy-like blooms atop a single,
branched stem. Summer-blooming, up to 3' tall.

Tragopogon dubius
Goatsbeard / Salsify

Asteraceae
Daisy Family

Non-Native and Invasive • Yellow

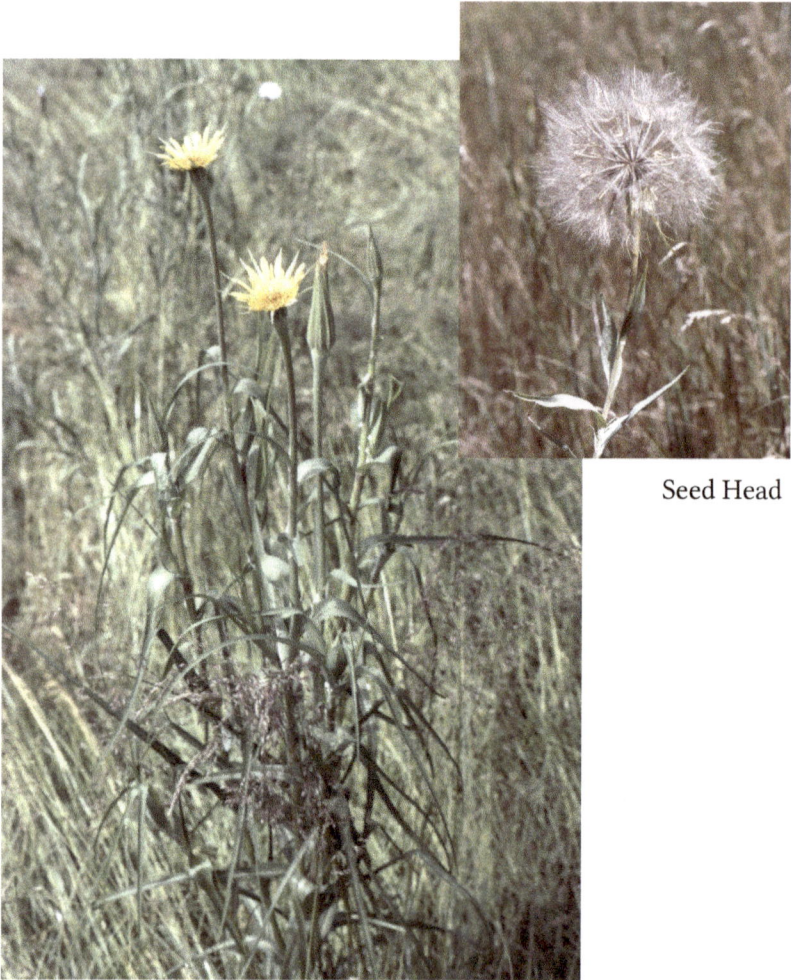

Seed Head

Photo by Frank Conley

Yellow bloom; looks like a giant dandelion when
gone to seed. Also called Oysterplant, as its
root supposedly tastes like oysters. 20-25" tall.

Berberis repens
Oregon Grape

Native • Yellow

Somewhat evergreen; many leaves
retained through winter. Yellow flowers
ripen into bluish berries. 12-20" tall.

Amsinckia menziesii
Fiddleneck

Boraginaceae
Forget-Me-Not Family

Native • Yellow

Flowers inconspicuous and coiled
like a scorpion's tail. 10-12" tall.

Anchusa officianalis
Common Bugloss / Blueweed

Boraginaceae
Forget-Me-Not Family

Non-Native and Invasive • Blue

Highly invasive weed, especially in moist
areas. Flower is deep blue. Listed as a
noxious weed in Oregon. Up to 3' tall.

Cynoglossum officinale   **Boraginaceae**
Houndstongue / Beggar Ticks   Forget-Me-Not Family

Non-Native and Invasive • Pink / Purple

Invasive biennial weed. First year plant is a rosette
of foliage with no bloom. Second year plant sends
up a blooming stalk with purple-red flowers that
bees love. Flowers ripen into velcro-like stickers. If
you don't find this one, your dog will! 15-30" tall.

Mertensia longifolia
Mountain Bluebell

Boraginaceae
Forget-Me-Not Family

Native • Blue

Blooms in late spring to early summer.
Very blue, tubular flowers. 8-12" tall.

Draba sp.
Common Mustard / Whitlow Grass

Native • White

This very large genus includes both native
and non-native species ranging in height
from 2-15". Many are small and not showy.

Erysimum capitatum
Wallflower

Brassicaceae
Mustard Family

Native • Yellow

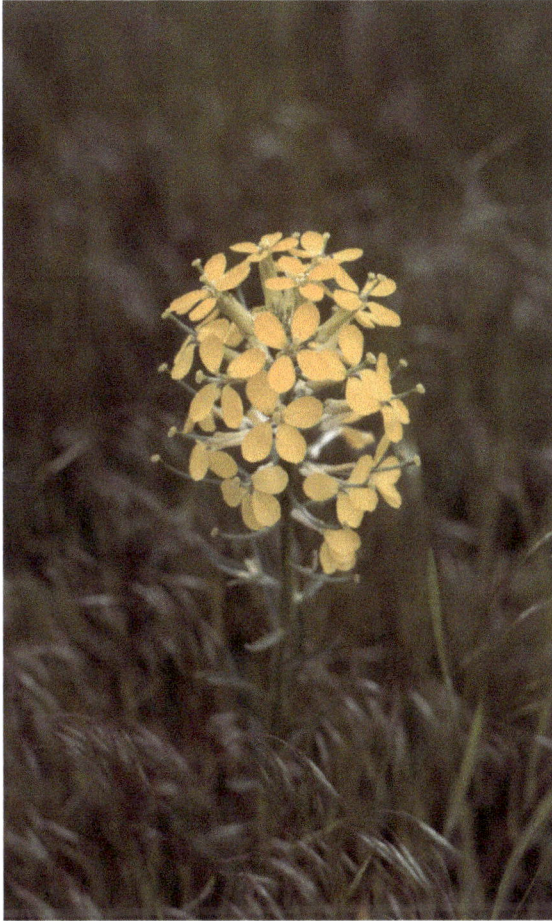

Yellow, sweet-smelling blooms appear in late spring and
early summer. Four petals per blossom is typical
of mustard family members. Around 12-15" tall.

Coryptantha missouriensis
Ball Cactus

Cactaceae
Cactus Family

Native • Pink

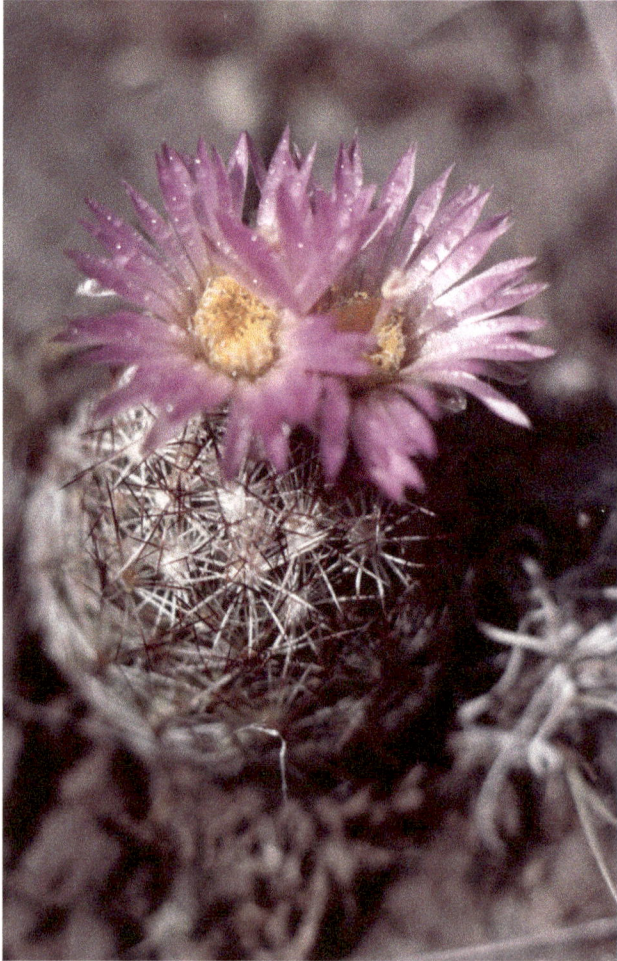

A rare cactus in Hells Canyon uplands. Very
compact and low to the ground. 2-3" tall.

Opuntia polycantha
Prickly Pear

Cactaceae
Cactus Family

Native • Yellow

The common cactus in Hells Canyon. Yellow
flowers and long, sharp spines. 10-15" tall.

Cerastium arvense   **Caryophyllaceae**
Chickweed   Pink Family

Native • White

Fairly conspicuous white flowers. Petals
are cleft on the ends. Plant grows 6-10" tall.

Eremogone aculeata
Needle-leaf Sandwort

Caryophyllaceae
Pink Family

Native • White

Short, cushiony clump of tiny white
blooms. Likes dry, gravelly soil. 6-10" tall.

Silene latifolia
Campion

Caryophyllaceae
Pink Family

Non-Native and Invasive • White

Showy white blossoms. Plant is 1-2' tall.

Convolvulus arvensis
Bindweed

Convolvulaceae
Morning Glory Family

Non-Native and Invasive • White / Pink

Photo by Frank Conley

Vining perennial invasive weed. The
vines can be up to 15" long but they cling
to the ground or climb up other plants.

Sedum stenopetalum   **Crassulaceae**
Sharp-leaved Sedum / Stonecrop   Stonecrop Family

Native • Yellow

Common in rocky, dry sites. Can also be
found among rocks in mountains. 3-10" tall.

Glossopetalon spinescens   **Crossosomataceae**
Snake R. Greasewood / Spiny Greenbush Brittlebush Family

Native • White

Distinctive green-stemmed shrub
with tiny leaves. Stands 1-3' tall.

Dipsacus fullonum
Teasel

Dipsacaceae
Teasel Family

Non-Native and Invasive • Purple

This tall biennial blooms purple, the
head ripening to a spiny brown seedhead
at the end of the stem. Up to 4' tall.

Euphorbia esula
Leafy Spurge

Euphorbiaceae
Spurge Family

Non-Native and Invasive • Yellow

Very invasive perennial weed with yellow-green
flowers and milky sap. Toxic to livestock. 1-3' tall.

Lupinus sp.
Lupine

Fabaceae
Pea Family

Native • Purple

The palmately compound leaves help identify
these plants, which consist of many different
species in many different habitats. 1-3' tall.

Trifolium macrocephalum
Big-headed Clover

Fabaceae
Pea Family

Native • Pink

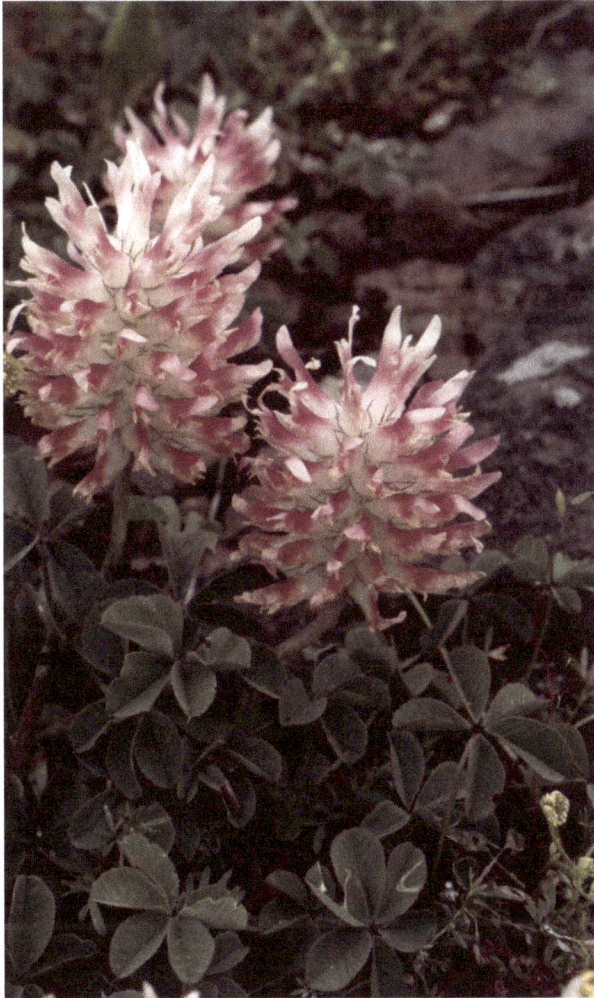

Flower heads are golf-ball sized
and 6-8" tall. Blooms in spring.

Frasera albicaulis
Shining Frasera

Gentianaceae
Gentian Family

Native • Purple

Can be confused with common camas, but
each flower has only four petals rather than
six, and the leaf edges are whitish. 6-12" tall.

Frasera speciosa
Green Gentian / Monument Plant

Gentianaceae
Gentian Family

Native • White

Biennial that forms a rosette of long, strap-like
leaves in its first year; sends up a 2-5' tall spike of
unique whitish-green flowers in its second year.

Erodium cicutarium
Filaree / Stork's Bill

Geraniaceae
Geranium Family

Non-Native and Invasive • Pink

Photo by Frank Conley

Photo by Frank Conley

Usually 3-4" tall with tiny pink flowers and dissected leaves. Flower matures into a long, pointed seed capsule which resembles a stork's bill to some.

Geranium viscosissimum
Sticky Geranium

Geraniaceae
Geranium Family

Native • Pink

Very showy, early-summer bloomer in
grasslands and open forests. 1-3' tall.

Philadelphus lewisii
Syringa / Mock Orange

Hydrangeaceae
Hydrangea Family

Native • White

A sweet-smelling shrub that blooms in
late spring in the canyon. Another of our
plants named for Captain Meriwether Lewis.

Hydrophyllum capitatum
Waterleaf / Woolly Breeches

Hydrophyllaceae
Waterleaf Family

Native • Purple

Ball-shaped flower cluster is often
hidden below leaves. Plant is 4-8" tall.

Phacelia hastata **Hydrophyllaceae**
Scorpion Weed Waterleaf Family

Native • White

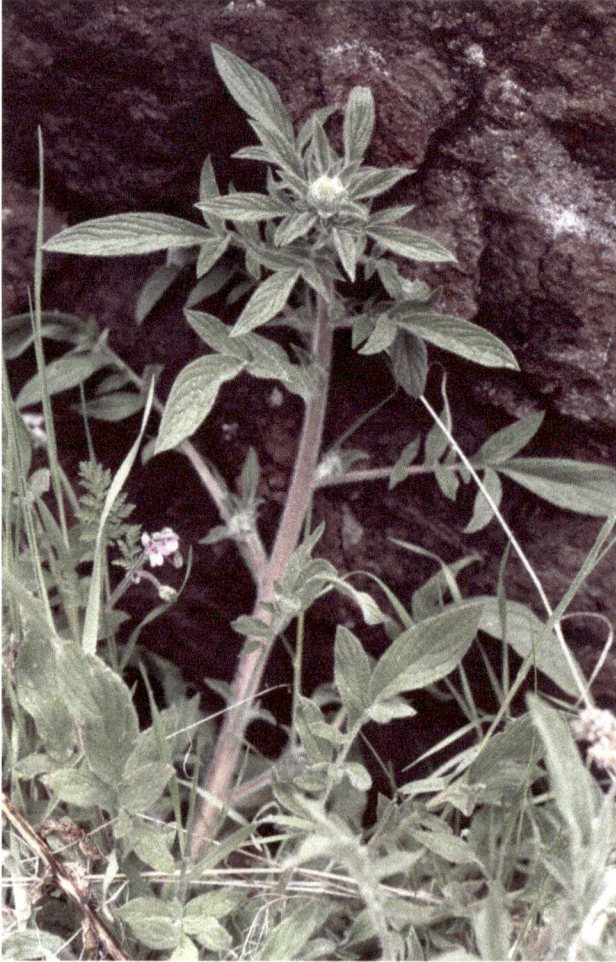

  Short-lived Oregon native perennial, so-named because
the flower spike is curled like a scorpion's tail. 12-20" tall.

Phacelia linearis
Linear-leaf Phacelia

Hydrophyllaceae
Waterleaf Family

Native • Blue

Light purplish-blue blossoms on a 6-10" tall plant.

Hypericum perforatum
St John's Wort

Hypericaceae
St. John's Wort Family

Non-Native and Invasive • Yellow

Invasive weed. Hold leaves up to light
and see the tiny 'perforations,' actually
lighter green markings, on leaves. 1-3' tall.

Sisyrinchium douglasii **Iridaceae**
Grass Widow Iris Family

Non-Native • Purple / Pink

Early-blooming, magenta-pink. 5-8" tall. Often
covers the ground with many plants together.

Agastache urticifolia
Horsemint

Lamiaceae
Mint Family

Native • Purple

This mint-family member grows in vigorous clumps
and is very popular with insect pollinators. Up to 3' tall.
Look for the square stem characteristic of the mint family.

Calochortus eurycarpus
Big-podded Mariposa

Liliaceae
Lily Family

Native • Purple / White

'Mariposa' means 'butterfly' in Spanish. These members
of the Lily family not only resemble butterflies, but
are also attractive nectar sources for them. 12-20" tall.

Fritillaria pudica
Yellowbell

Liliaceae
Lily Family

Native • Yellow

One of the earliest spring blooms. 3-6"
tall. Forms a round white seed pod.

Linum perenne
Blue Flax

Linaceae
Flax Family

Native • Blue

A showy yet delicate bloom on the end
of a 15-24" stem. Leaves are very small.

Mentzerlia laevicaulus
Blazing Star

Loasaceae
Loasa Family

Native • Yellow

Late-summer bloom. Very branching, 2-3' tall.

Toxicoscordion paniculatum
Panicled Death Camas

Melanthiaceae
Bunchflower Family

Native • White

Although the blooms are completely different than those
of common camas, the roots are very similar in looks
but are toxic. Definitely not used as food. 12-15" tall.

Claytonia megarhiza
Spring Beauty

Montiaceae
Spring Beauty Family

Native • Pink / White

As its name implies, this small plant blooms
in spring. All parts are edible. 6-10" tall.

Lewisia columbiana
Columbia Lewisia

Montiaceae
Spring Beauty Family

Native • White

A semi-succulent plant named after Meriwether
Lewis of the Lewis and Clark expedition.
Only found in dry, rocky sites. 6-10" tall.

Mirabilis mcfarlanei **Nyctaginaceae**
McFarlane's Four o'clock Four o'clock Family

Native • Pink

A very distinctive, round-leaved perennial known from
only a few upland sites in Hells Canyon. Grows to
2' tall. This plant is listed as Endangered under the
Endangered Species Act, making it illegal to pick or
collect seeds. Named after an early Snake River boatman.

Clarkia pulchella
Clarkia / Farewell to Spring

Onagraceae
Evening Primrose Family

Native • Pink

This plant was one of many that were collected by Lewis and Clark, and named accordingly. 12-15" tall annual.

Oenothera caespitosa
Desert Evening Primrose

Onagraceae
Evening Primrose Family

Native • White

Four heart-shaped white petals per blossom. 15-20" tall.

Epipactis gigantea
Chatterbox Orchid / Giant Helleborine

Orchidaceae
Orchid Family

Native • Green

One of more than a dozen species of orchids
in northeast Oregon, this one is found along
moist springs in Hells Canyon. Up to 3' tall.

Castilleja sp.
Paintbrush

Orobanchaceae
Paintbrush Family

Native • Red / White / Yellow

We have many varieties of paintbrush, in colors ranging
from pale yellow to orangeish-red to scarlet. All are
partially parasitic plants, often associated with sagebrush.

Orobanche uniflora
Broomrape

Orobanchaceae
Paintbrush Family

Native • Purple

This weirdly-named plant is usually
found growing singly. It is 3-5" tall and
parasitic on other plants, because it has no
chlorophyll to manufacture its own food.

Orthocarpus tenuifolius
Owl Clover

Orobanchaceae
Paintbrush Family

Native • Yellow / Pink

The taxonomy of this plant is confusing and often
being revised. It is a close relative of paintbrush.
Formerly in family Scrophulariaceae. 6-12" tall.

Paeonia brownii
Brown's Peony

Paeoniaceae
Peony Family

Native • Brown

The unique brownish bloom and much-lobed,
gray-green leaves identify this plant. Not nearly
as showy as garden-variety peonies. 10-16" tall.

Dicentra cucullaria
Dutchman's Breeches

Papaveraceae
Poppy Family

Native • White

Unmistakable blooms on long, drooping
stems. Always in moist areas. 10-15" tall.

Collinsia parviflora
Blue-eyed Mary

Plantaginaceae
Plantain Family

Native • Blue

A tiny spring-blooming plant that
seldom exceeds 4 inches in height.

Linaria vulgaris
Butter and Eggs Toadflax

Plantaginaceae
Plantain Family

Non-Native and Invasive • Yellow

An invasive weed resembling garden snapdragons.
May be harmful to livestock. 10-20" tall.

Penstemon deustus
Hot Rock or Scabland Penstemon

Plantaginaceae
Plantain Family

Native • White

As its name implies, this white penstemon
prefers hot, dry, rocky sites. 10-20" tall.

Penstemon fruticosus
Shrubby Penstemon

Plantaginaceae
Plantain Family

Native • Purple

A very showy semi-shrub, often
seen on rocky cliffs. 2-3' tall.

Penstemon sp.
Penstemon

Plantaginaceae
Plantain Family

Native • Blue / Purple

Another large genus. Most of ours are blue,
purple, or white, and all have showy spikes
of summer-blooming flowers. 10-20" tall.

Ipomopsis aggregata
Scarlet Gilia / Skyrocket

Polemoniaceae
Phlox Family

Native • Red

Hummingbirds love this flower, which comes
in shades from light pink to scarlet. 20-24" tall.

Phlox longifolia
Long-leaved Phlox

Polemoniaceae
Phlox Family

Native • Pink

A common spring bloom in the grasslands
and along the Snake River. 6-10" tall.

Eriogonum heracleiodes
Creamy Buckwheat

Polygonaceae
Buckwheat Family

Native • White

Blooms in summer. Flowers are in groups
called umbels at the ends of 12-15" stems.
Foliage is grayish green and mostly basal.

96

Eriogonum thymoides
Thyme-leaved Buckwheat

Polygonaceae
Buckwheat Family

Native • Red / Yellow

Thyme-leaved buckwheat and creamy buckwheat
form semi-shrubby mats 8-15" tall on dry slopes.

Rumex acetosella
Sheep Sorrel

Polygonaceae
Buckwheat Family

Non-Native and Invasive • Red

This plant is a good source of
Vitamin C. Grows 6-12" tall.

Dodecatheon pulchellum
Shooting Star

Primulaceae
Primrose Family

Native • Pink

An early-spring bloomer. Leafless stem above
a basal rosette of leaves. Usually 5-10" tall.

Clematis hirsutissima   **Ranunculaceae**
Sugar Scoop / Vase Flower   Buttercup Family

Native • Purple

Seed Head

Many of the members of the buttercup family look
nothing like a typical 'buttercup.' This is a good example.
Seed heads resemble little dust mops. 12-20" tall.

Delphinium sp.
Larkspur

Ranunculaceae
Buttercup Family

Native • Blue

This native plant is toxic to cattle if eaten in quantity.
We have several common species. Grows 10-24" tall.

Ranunculus glaberrimus
Sage Buttercup

Ranunculaceae
Buttercup Family

Native • Yellow

This is one of the first flowers to show
up in spring. It is often hidden in the
brown grass of winter. Very small, 2-5" tall.

Amelanchier alnifolia
Serviceberry / Sarvisberry

Rosaceae
Rose Family

Native • White

The first shrub to bloom in spring. Its fruits are dark blue, small and seedy, but good to eat. Up to 10' tall.

Crataegus columbiana
Hawthorn

Rosaceae
Rose Family

Native • White

A shrub that often forms dense thickets. Foliage turns red in fall. Long, straight thorns on stems. Up to 12' tall.

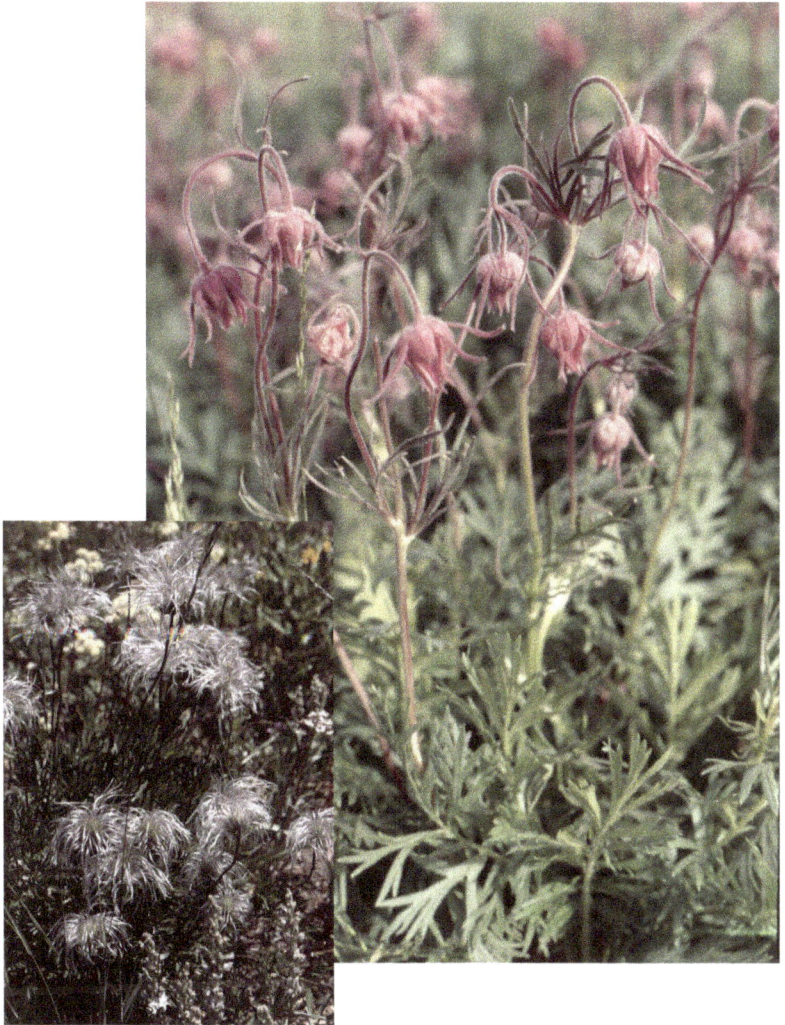

Geum triflorum
Prairie Smoke / Old Man's Whiskers

Rosaceae
Rose Family

Native • Pink

Seed Head

Very common on grassy slopes and meadows.
The name comes from the way extensive
patches look when in seed. 12-15" tall.

Holodiscus discolor
Ocean Spray

Rosaceae
Rose Family

Native • White

This shrub is found in partial shade, in pine or
Douglas-fir stands. It blooms in mid-summer. 6-8' tall.

Physocarpus malvaceus
Ninebark

Rosaceae
Rose Family

Native • White

This low-growing shrub was used by the
Nez Perce for its fiber in basket-making.
It turns red in the fall. Up to 4' tall.

Potentilla sp.
Cinquefoil

Rosaceae
Rose Family

Native • Yellow

This large genus is well-represented in Hells Canyon.
Besides several native species, there is a non-native,
invasive variety, *Potentilla recta* or Sulfur Cinquefoil,
that has very hairy stems and very pale yellow flowers.

Prunus virginiana
Chokecherry

Rosaceae
Rose Family

Native • White

A tall shrub, usually found close to waterways.
Clusters of bright red berries in seed. Birds
and bears love them. This was an important
food plant for the Nez Perce. Grows to 15' tall.

Rosa sp.
Wild Rose

Rosaceae
Rose Family

Native • Pink

Rose hips, the fruit of this plant, are high in
Vitamin C. We have several native species as
well as non-native invasives. Usually 2-4' tall.

Rubus sp.
Raspberry

Rosaceae
Rose Family

Native • White

This native does not form big thickets as
blackberries do. The berries are delicious. 2-3' tall.

Heuchera cylindrica
Alum Root

Saxifragaceae
Saxifrage Family

Native • White

This plant's root was used medicinally by
the Nez Perce and by settlers. 12-20" tall.

Lithophragma parviflorum
Prairie Star

Saxifragaceae
Saxifrage Family

Native • White

This delicate little plant's genus is derived
from the Latin for 'rock breaker.' Apparently
it is tougher than it looks. It grows 4-6" tall.

Verbascum thapsus
Mullein

Scrophulariaceae
Figwort Family

Non-Native and Invasive • Yellow

Photo by Jenner Hanni

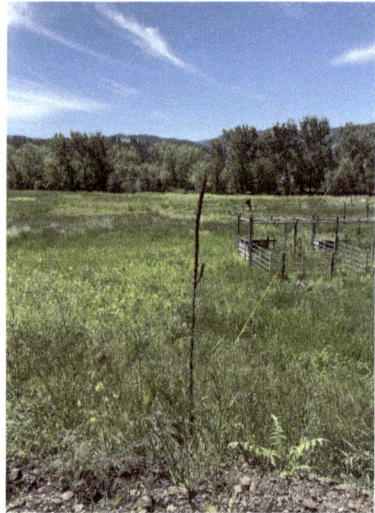

The fuzzy, thick leaves of this yellow-flowered
plant are distinctive. It grows 4-6' tall.

Urtica dioecea
Stinging Nettle

Urticaceae
Nettle Family

Native • White

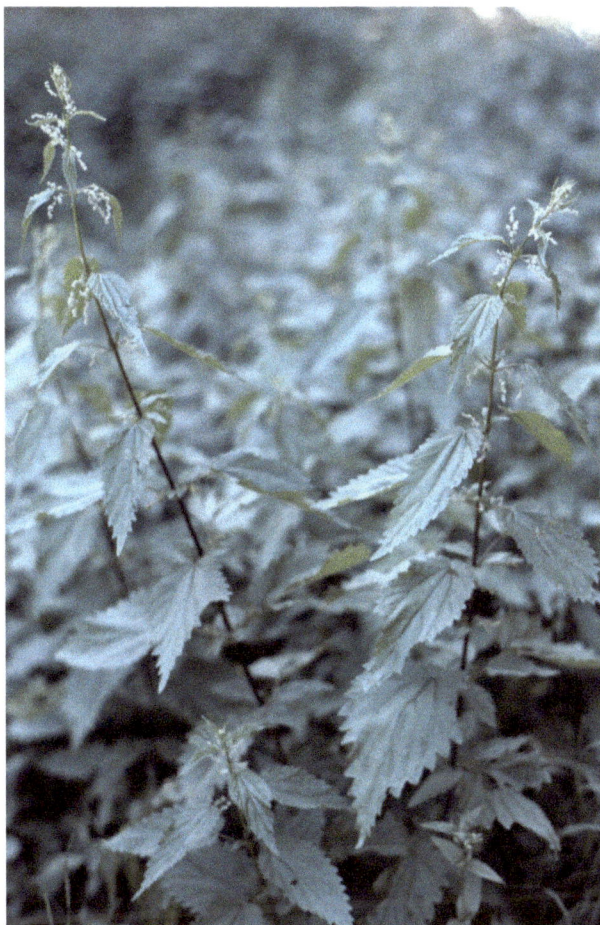

This invasive plant is covered with tiny hairs
that irritate skin. Young plants can be carefully
gathered and boiled for a spring green. 3-4' tall.

Viola adunca
Long-spurred Violet

Violaceae
Violet Family

Native • Purple

This small purple bloom has a distinct spur
visible behind the main petals. 4-8" tall.

Viola purpurea
Mountain Yellow Violet

Violaceae
Violet Family

Native • Yellow

A yellow version of the common
purple-petaled violet. 4-8" tall.

Index by Common Name

Index by Color

Purple